DETAILS

NAME	
CONTACT NUMBER	
EMAIL	
COMPANY	

EMERGENCY DETAILS:

NAME:	NAME:
CONTACT NUMBER:	CONTACT NUMBER:

IMPORTANT: ALWAYS APPEND CONTRACT/JOB NUMBER ON EVERY FOLLOWING PAGE WITH THE DAY'S DATE, FOR EXAMPLE:
JOBNAME-SITELOG-2019-06-27

DATE:	/	/	DAY:	M	T	W	T	F	S	Su
FOREMAN:										
CONTRACT No.:										

WEATHER CONDITIONS:		VISITORS:
AM	PM	
HOURS LOST DUE TO BAD WEATHER:		

SCHEDULE:		PROBLEMS/DELAYS:
Completion Date		
Days Ahead of Schedule		
Days Behind Schedule		

SAFETY ISSUES:	ACCIDENTS/INCIDENTS:

SUMMARY OF WORK PERFORMED TODAY

SIGNATURE	NAME

EMPLOYEE/ CONTRACTOR	TRADE	CONTRACTED HOURS	OVERTIME

EQUIPMENT ON SITE	NO. UNITS	Working	
		Yes	No

MATERIALS DELIVERED	NO. UNITS	EQUIPMENT RENTED	FROM & RATE

NOTES

DATE:	/	/	DAY:	M	T	W	T	F	S	Su
FOREMAN:										
CONTRACT No.:										

WEATHER CONDITIONS:		VISITORS:
AM	**PM**	
HOURS LOST DUE TO BAD WEATHER:		

SCHEDULE:		PROBLEMS/DELAYS:
Completion Date		
Days Ahead of Schedule		
Days Behind Schedule		

SAFETY ISSUES:	ACCIDENTS/INCIDENTS:

SUMMARY OF WORK PERFORMED TODAY

SIGNATURE	NAME

EMPLOYEE/ CONTRACTOR	TRADE	CONTRACTED HOURS	OVERTIME

EQUIPMENT ON SITE	NO. UNITS	Working	
		Yes	No

MATERIALS DELIVERED	NO. UNITS	EQUIPMENT RENTED	FROM & RATE

NOTES

DATE: / /	DAY: M T W T F S Su
FOREMAN:	
CONTRACT No.:	

WEATHER CONDITIONS:		VISITORS:
AM	PM	
HOURS LOST DUE TO BAD WEATHER:		

SCHEDULE:		PROBLEMS/DELAYS:
Completion Date		
Days Ahead of Schedule		
Days Behind Schedule		

SAFETY ISSUES:	ACCIDENTS/INCIDENTS:

SUMMARY OF WORK PERFORMED TODAY

SIGNATURE	NAME

EMPLOYEE/ CONTRACTOR	TRADE	CONTRACTED HOURS	OVERTIME

EQUIPMENT ON SITE	NO. UNITS	Working	
		Yes	No

MATERIALS DELIVERED	NO. UNITS	EQUIPMENT RENTED	FROM & RATE

NOTES

_____

_____

_____

_____

DATE: / /		DAY:	M	T	W	T	F	S	Su
FOREMAN:									
CONTRACT No.:									

WEATHER CONDITIONS:		VISITORS:
AM	**PM**	
HOURS LOST DUE TO BAD WEATHER:		

SCHEDULE:		PROBLEMS/DELAYS:
Completion Date		
Days Ahead of Schedule		
Days Behind Schedule		

SAFETY ISSUES:	ACCIDENTS/INCIDENTS:

SUMMARY OF WORK PERFORMED TODAY

SIGNATURE	NAME

EMPLOYEE/ CONTRACTOR	TRADE	CONTRACTED HOURS	OVERTIME

EQUIPMENT ON SITE	NO. UNITS	Working	
		Yes	No

MATERIALS DELIVERED	NO. UNITS	EQUIPMENT RENTED	FROM & RATE

NOTES

_____
_____
_____
_____
_____

DATE:	/	/	DAY:	M	T	W	T	F	S	Su
FOREMAN:										
CONTRACT No.:										

WEATHER CONDITIONS:		VISITORS:
AM	**PM**	
HOURS LOST DUE TO BAD WEATHER:		

SCHEDULE:		PROBLEMS/DELAYS:
Completion Date		
Days Ahead of Schedule		
Days Behind Schedule		

SAFETY ISSUES:	ACCIDENTS/INCIDENTS:

SUMMARY OF WORK PERFORMED TODAY

SIGNATURE	NAME

EMPLOYEE/ CONTRACTOR	TRADE	CONTRACTED HOURS	OVERTIME

EQUIPMENT ON SITE	NO. UNITS	Working	
		Yes	No

MATERIALS DELIVERED	NO. UNITS	EQUIPMENT RENTED	FROM & RATE

NOTES

DATE: / /	DAY: M T W T F S Su
FOREMAN:	
CONTRACT No.:	

WEATHER CONDITIONS:		VISITORS:
AM	PM	
HOURS LOST DUE TO BAD WEATHER:		

SCHEDULE:		PROBLEMS/DELAYS:
Completion Date		
Days Ahead of Schedule		
Days Behind Schedule		

SAFETY ISSUES:	ACCIDENTS/INCIDENTS:

SUMMARY OF WORK PERFORMED TODAY

SIGNATURE	NAME

EMPLOYEE/ CONTRACTOR	TRADE	CONTRACTED HOURS	OVERTIME

EQUIPMENT ON SITE	NO. UNITS	Working	
		Yes	No

MATERIALS DELIVERED	NO. UNITS	EQUIPMENT RENTED	FROM & RATE

NOTES

DATE: / /	DAY: M T W T F S Su
FOREMAN:	
CONTRACT No.:	

WEATHER CONDITIONS:		VISITORS:
AM	PM	
HOURS LOST DUE TO BAD WEATHER:		

SCHEDULE:		PROBLEMS/DELAYS:
Completion Date		
Days Ahead of Schedule		
Days Behind Schedule		

SAFETY ISSUES:	ACCIDENTS/INCIDENTS:

SUMMARY OF WORK PERFORMED TODAY

SIGNATURE	NAME

EMPLOYEE/ CONTRACTOR	TRADE	CONTRACTED HOURS	OVERTIME

EQUIPMENT ON SITE	NO. UNITS	Working	
		Yes	No

MATERIALS DELIVERED	NO. UNITS	EQUIPMENT RENTED	FROM & RATE

NOTES

_____

_____

_____

_____

DATE: / /	DAY: M T W T F S Su
FOREMAN:	
CONTRACT No.:	

WEATHER CONDITIONS:		VISITORS:
AM	**PM**	
HOURS LOST DUE TO BAD WEATHER:		

SCHEDULE:		PROBLEMS/DELAYS:
Completion Date		
Days Ahead of Schedule		
Days Behind Schedule		

SAFETY ISSUES:	ACCIDENTS/INCIDENTS:

SUMMARY OF WORK PERFORMED TODAY

SIGNATURE	NAME

EMPLOYEE/ CONTRACTOR	TRADE	CONTRACTED HOURS	OVERTIME

EQUIPMENT ON SITE	NO. UNITS	Working	
		Yes	No

MATERIALS DELIVERED	NO. UNITS	EQUIPMENT RENTED	FROM & RATE

NOTES

_____

_____

_____

_____

DATE: / /	DAY: M T W T F S Su
FOREMAN:	
CONTRACT No.:	

WEATHER CONDITIONS:		VISITORS:
AM	**PM**	
HOURS LOST DUE TO BAD WEATHER:		

SCHEDULE:		PROBLEMS/DELAYS:
Completion Date		
Days Ahead of Schedule		
Days Behind Schedule		

SAFETY ISSUES:	ACCIDENTS/INCIDENTS:

SUMMARY OF WORK PERFORMED TODAY

SIGNATURE	NAME

EMPLOYEE/ CONTRACTOR	TRADE	CONTRACTED HOURS	OVERTIME

EQUIPMENT ON SITE	NO. UNITS	Working	
		Yes	No

MATERIALS DELIVERED	NO. UNITS	EQUIPMENT RENTED	FROM & RATE

NOTES

_____

_____

_____

_____

_____

DATE:	/	/	DAY:	M	T	W	T	F	S	Su
FOREMAN:										
CONTRACT No.:										

WEATHER CONDITIONS:		VISITORS:
AM	**PM**	
HOURS LOST DUE TO BAD WEATHER:		

SCHEDULE:		PROBLEMS/DELAYS:
Completion Date		
Days Ahead of Schedule		
Days Behind Schedule		

SAFETY ISSUES:	ACCIDENTS/INCIDENTS:

SUMMARY OF WORK PERFORMED TODAY

SIGNATURE	NAME

EMPLOYEE/ CONTRACTOR	TRADE	CONTRACTED HOURS	OVERTIME

EQUIPMENT ON SITE	NO. UNITS	Working	
		Yes	No

MATERIALS DELIVERED	NO. UNITS	EQUIPMENT RENTED	FROM & RATE

NOTES

DATE: / /	DAY: M T W T F S Su
FOREMAN:	
CONTRACT No.:	

WEATHER CONDITIONS:		VISITORS:
AM	**PM**	
HOURS LOST DUE TO BAD WEATHER:		

SCHEDULE:		PROBLEMS/DELAYS:
Completion Date		
Days Ahead of Schedule		
Days Behind Schedule		

SAFETY ISSUES:	ACCIDENTS/INCIDENTS:

SUMMARY OF WORK PERFORMED TODAY

SIGNATURE	NAME

EMPLOYEE/ CONTRACTOR	TRADE	CONTRACTED HOURS	OVERTIME

EQUIPMENT ON SITE	NO. UNITS	Working	
		Yes	No

MATERIALS DELIVERED	NO. UNITS	EQUIPMENT RENTED	FROM & RATE

NOTES

DATE: / /	DAY: M T W T F S Su

FOREMAN:	

CONTRACT No.:	

WEATHER CONDITIONS:		VISITORS:
AM	PM	
HOURS LOST DUE TO BAD WEATHER:		

SCHEDULE:		PROBLEMS/DELAYS:
Completion Date		
Days Ahead of Schedule		
Days Behind Schedule		

SAFETY ISSUES:	ACCIDENTS/INCIDENTS:

SUMMARY OF WORK PERFORMED TODAY

SIGNATURE	NAME

EMPLOYEE/ CONTRACTOR	TRADE	CONTRACTED HOURS	OVERTIME

EQUIPMENT ON SITE	NO. UNITS	Working	
		Yes	No

MATERIALS DELIVERED	NO. UNITS	EQUIPMENT RENTED	FROM & RATE

NOTES

DATE: / /	DAY: M T W T F S Su
FOREMAN:	
CONTRACT No.:	

WEATHER CONDITIONS:		VISITORS:
AM	**PM**	
HOURS LOST DUE TO BAD WEATHER:		

SCHEDULE:		PROBLEMS/DELAYS:
Completion Date		
Days Ahead of Schedule		
Days Behind Schedule		

SAFETY ISSUES:	ACCIDENTS/INCIDENTS:

SUMMARY OF WORK PERFORMED TODAY

SIGNATURE	NAME

EMPLOYEE/ CONTRACTOR	TRADE	CONTRACTED HOURS	OVERTIME

EQUIPMENT ON SITE	NO. UNITS	Working	
		Yes	No

MATERIALS DELIVERED	NO. UNITS	EQUIPMENT RENTED	FROM & RATE

NOTES

DATE: / /			DAY: M T W T F S Su
FOREMAN:			
CONTRACT No.:			

WEATHER CONDITIONS:		VISITORS:
AM	**PM**	
HOURS LOST DUE TO BAD WEATHER:		

SCHEDULE:		PROBLEMS/DELAYS:
Completion Date		
Days Ahead of Schedule		
Days Behind Schedule		

SAFETY ISSUES:	ACCIDENTS/INCIDENTS:

SUMMARY OF WORK PERFORMED TODAY

SIGNATURE	NAME

EMPLOYEE/ CONTRACTOR	TRADE	CONTRACTED HOURS	OVERTIME

EQUIPMENT ON SITE	NO. UNITS	Working	
		Yes	No

MATERIALS DELIVERED	NO. UNITS	EQUIPMENT RENTED	FROM & RATE

NOTES

_____
_____
_____
_____

DATE: / /	DAY: M T W T F S Su
FOREMAN:	
CONTRACT No.:	

WEATHER CONDITIONS:		VISITORS:
AM	**PM**	
HOURS LOST DUE TO BAD WEATHER:		

SCHEDULE:		PROBLEMS/DELAYS:
Completion Date		
Days Ahead of Schedule		
Days Behind Schedule		

SAFETY ISSUES:	ACCIDENTS/INCIDENTS:

SUMMARY OF WORK PERFORMED TODAY

SIGNATURE	NAME

EMPLOYEE/ CONTRACTOR	TRADE	CONTRACTED HOURS	OVERTIME

EQUIPMENT ON SITE	NO. UNITS	Working	
		Yes	No

MATERIALS DELIVERED	NO. UNITS	EQUIPMENT RENTED	FROM & RATE

NOTES

_____

_____

_____

_____

_____

DATE: / /	DAY: M T W T F S Su
FOREMAN:	
CONTRACT No.:	

WEATHER CONDITIONS:		VISITORS:
AM	**PM**	
HOURS LOST DUE TO BAD WEATHER:		

SCHEDULE:		PROBLEMS/DELAYS:
Completion Date		
Days Ahead of Schedule		
Days Behind Schedule		

SAFETY ISSUES:	ACCIDENTS/INCIDENTS:

SUMMARY OF WORK PERFORMED TODAY

SIGNATURE	NAME

EMPLOYEE/ CONTRACTOR	TRADE	CONTRACTED HOURS	OVERTIME

EQUIPMENT ON SITE	NO. UNITS	Working	
		Yes	No

MATERIALS DELIVERED	NO. UNITS	EQUIPMENT RENTED	FROM & RATE

NOTES

_____
_____
_____
_____

DATE: / /	DAY: M T W T F S Su
FOREMAN:	
CONTRACT No.:	

WEATHER CONDITIONS:		VISITORS:
AM	**PM**	
HOURS LOST DUE TO BAD WEATHER:		

SCHEDULE:		PROBLEMS/DELAYS:
Completion Date		
Days Ahead of Schedule		
Days Behind Schedule		

SAFETY ISSUES:	ACCIDENTS/INCIDENTS:

SUMMARY OF WORK PERFORMED TODAY

SIGNATURE	NAME

EMPLOYEE/ CONTRACTOR	TRADE	CONTRACTED HOURS	OVERTIME

EQUIPMENT ON SITE	NO. UNITS	Working	
		Yes	No

MATERIALS DELIVERED	NO. UNITS	EQUIPMENT RENTED	FROM & RATE

NOTES

_____

_____

_____

_____

DATE:	/	/	DAY:	M	T	W	T	F	S	Su
FOREMAN:										
CONTRACT No.:										

WEATHER CONDITIONS:		VISITORS:
AM	**PM**	
HOURS LOST DUE TO BAD WEATHER:		

SCHEDULE:		PROBLEMS/DELAYS:
Completion Date		
Days Ahead of Schedule		
Days Behind Schedule		

SAFETY ISSUES:	ACCIDENTS/INCIDENTS:

SUMMARY OF WORK PERFORMED TODAY

SIGNATURE	NAME

EMPLOYEE/ CONTRACTOR	TRADE	CONTRACTED HOURS	OVERTIME

EQUIPMENT ON SITE	NO. UNITS	Working	
		Yes	No

MATERIALS DELIVERED	NO. UNITS	EQUIPMENT RENTED	FROM & RATE

NOTES

_____

_____

_____

_____

_____

DATE: / /	DAY: M T W T F S Su
FOREMAN:	
CONTRACT No.:	

WEATHER CONDITIONS:		VISITORS:
AM	**PM**	
HOURS LOST DUE TO BAD WEATHER:		

SCHEDULE:		PROBLEMS/DELAYS:
Completion Date		
Days Ahead of Schedule		
Days Behind Schedule		

SAFETY ISSUES:	ACCIDENTS/INCIDENTS:

SUMMARY OF WORK PERFORMED TODAY

SIGNATURE	NAME

EMPLOYEE/ CONTRACTOR	TRADE	CONTRACTED HOURS	OVERTIME

EQUIPMENT ON SITE	NO. UNITS	Working	
		Yes	No

MATERIALS DELIVERED	NO. UNITS	EQUIPMENT RENTED	FROM & RATE

NOTES

_____

_____

_____

_____

DATE: / /	DAY: M T W T F S Su
FOREMAN:	
CONTRACT No.:	

WEATHER CONDITIONS:		VISITORS:
AM	**PM**	
HOURS LOST DUE TO BAD WEATHER:		

SCHEDULE:		PROBLEMS/DELAYS:
Completion Date		
Days Ahead of Schedule		
Days Behind Schedule		

SAFETY ISSUES:	ACCIDENTS/INCIDENTS:

SUMMARY OF WORK PERFORMED TODAY

SIGNATURE	NAME

EMPLOYEE/ CONTRACTOR	TRADE	CONTRACTED HOURS	OVERTIME

EQUIPMENT ON SITE	NO. UNITS	Working	
		Yes	No

MATERIALS DELIVERED	NO. UNITS	EQUIPMENT RENTED	FROM & RATE

NOTES

DATE: / /	DAY: M T W T F S Su
FOREMAN:	
CONTRACT No.:	

WEATHER CONDITIONS:		VISITORS:
AM	**PM**	
HOURS LOST DUE TO BAD WEATHER:		

SCHEDULE:		PROBLEMS/DELAYS:
Completion Date		
Days Ahead of Schedule		
Days Behind Schedule		

SAFETY ISSUES:	ACCIDENTS/INCIDENTS:

SUMMARY OF WORK PERFORMED TODAY

SIGNATURE	NAME

EMPLOYEE/ CONTRACTOR	TRADE	CONTRACTED HOURS	OVERTIME

EQUIPMENT ON SITE	NO. UNITS	Working	
		Yes	No

MATERIALS DELIVERED	NO. UNITS	EQUIPMENT RENTED	FROM & RATE

NOTES

DATE: / /	DAY: M T W T F S Su
FOREMAN:	
CONTRACT No.:	

WEATHER CONDITIONS:		VISITORS:
AM	**PM**	
HOURS LOST DUE TO BAD WEATHER:		

SCHEDULE:		PROBLEMS/DELAYS:
Completion Date		
Days Ahead of Schedule		
Days Behind Schedule		

SAFETY ISSUES:	ACCIDENTS/INCIDENTS:

SUMMARY OF WORK PERFORMED TODAY

SIGNATURE	NAME

EMPLOYEE/ CONTRACTOR	TRADE	CONTRACTED HOURS	OVERTIME

EQUIPMENT ON SITE	NO. UNITS	Working	
		Yes	No

MATERIALS DELIVERED	NO. UNITS	EQUIPMENT RENTED	FROM & RATE

NOTES

DATE:	/	/	DAY:	M	T	W	T	F	S	Su
FOREMAN:										
CONTRACT No.:										

WEATHER CONDITIONS:		VISITORS:
AM	**PM**	
HOURS LOST DUE TO BAD WEATHER:		

SCHEDULE:		PROBLEMS/DELAYS:
Completion Date		
Days Ahead of Schedule		
Days Behind Schedule		

SAFETY ISSUES:	ACCIDENTS/INCIDENTS:

SUMMARY OF WORK PERFORMED TODAY

SIGNATURE	NAME

EMPLOYEE/ CONTRACTOR	TRADE	CONTRACTED HOURS	OVERTIME

EQUIPMENT ON SITE	NO. UNITS	Working	
		Yes	No

MATERIALS DELIVERED	NO. UNITS	EQUIPMENT RENTED	FROM & RATE

NOTES

_____

_____

_____

_____

DATE: / /	DAY:	M	T	W	T	F	S	Su
FOREMAN:								
CONTRACT No.:								

WEATHER CONDITIONS:		VISITORS:
AM	**PM**	
HOURS LOST DUE TO BAD WEATHER:		

SCHEDULE:		PROBLEMS/DELAYS:
Completion Date		
Days Ahead of Schedule		
Days Behind Schedule		

SAFETY ISSUES:	ACCIDENTS/INCIDENTS:

SUMMARY OF WORK PERFORMED TODAY

SIGNATURE	NAME

EMPLOYEE/ CONTRACTOR	TRADE	CONTRACTED HOURS	OVERTIME

EQUIPMENT ON SITE	NO. UNITS	Working	
		Yes	No

MATERIALS DELIVERED	NO. UNITS	EQUIPMENT RENTED	FROM & RATE

NOTES

_____
_____
_____
_____
_____

DATE: / /	DAY: M T W T F S Su
FOREMAN:	
CONTRACT No.:	

WEATHER CONDITIONS:		VISITORS:
AM	PM	
HOURS LOST DUE TO BAD WEATHER:		

SCHEDULE:		PROBLEMS/DELAYS:
Completion Date		
Days Ahead of Schedule		
Days Behind Schedule		

SAFETY ISSUES:	ACCIDENTS/INCIDENTS:

SUMMARY OF WORK PERFORMED TODAY

SIGNATURE	NAME

EMPLOYEE/ CONTRACTOR	TRADE	CONTRACTED HOURS	OVERTIME

EQUIPMENT ON SITE	NO. UNITS	Working	
		Yes	No

MATERIALS DELIVERED	NO. UNITS	EQUIPMENT RENTED	FROM & RATE

NOTES

_____

_____

_____

_____

DATE: / /	DAY: M T W T F S Su
FOREMAN:	
CONTRACT No.:	

WEATHER CONDITIONS:		VISITORS:
AM	**PM**	
HOURS LOST DUE TO BAD WEATHER:		

SCHEDULE:		PROBLEMS/DELAYS:
Completion Date		
Days Ahead of Schedule		
Days Behind Schedule		

SAFETY ISSUES:	ACCIDENTS/INCIDENTS:

SUMMARY OF WORK PERFORMED TODAY

SIGNATURE	NAME

EMPLOYEE/ CONTRACTOR	TRADE	CONTRACTED HOURS	OVERTIME

EQUIPMENT ON SITE	NO. UNITS	Working	
		Yes	No

MATERIALS DELIVERED	NO. UNITS	EQUIPMENT RENTED	FROM & RATE

NOTES

_____

_____

_____

_____

_____

DATE: / /	DAY:	M	T	W	T	F	S	Su
FOREMAN:								
CONTRACT No.:								

WEATHER CONDITIONS:		VISITORS:
AM	PM	
HOURS LOST DUE TO BAD WEATHER:		

SCHEDULE:		PROBLEMS/DELAYS:
Completion Date		
Days Ahead of Schedule		
Days Behind Schedule		

SAFETY ISSUES:	ACCIDENTS/INCIDENTS:

SUMMARY OF WORK PERFORMED TODAY

SIGNATURE	NAME

EMPLOYEE/ CONTRACTOR	TRADE	CONTRACTED HOURS	OVERTIME

EQUIPMENT ON SITE	NO. UNITS	Working	
		Yes	No

MATERIALS DELIVERED	NO. UNITS	EQUIPMENT RENTED	FROM & RATE

NOTES

_____

_____

_____

_____

_____

DATE: / /	DAY: M T W T F S Su
FOREMAN:	
CONTRACT No.:	

WEATHER CONDITIONS:		VISITORS:
AM	PM	
HOURS LOST DUE TO BAD WEATHER:		

SCHEDULE:		PROBLEMS/DELAYS:
Completion Date		
Days Ahead of Schedule		
Days Behind Schedule		

SAFETY ISSUES:	ACCIDENTS/INCIDENTS:

SUMMARY OF WORK PERFORMED TODAY

SIGNATURE	NAME

EMPLOYEE/ CONTRACTOR	TRADE	CONTRACTED HOURS	OVERTIME

EQUIPMENT ON SITE	NO. UNITS	Working	
		Yes	No

MATERIALS DELIVERED	NO. UNITS	EQUIPMENT RENTED	FROM & RATE

NOTES

_____
_____
_____
_____

DATE: / /	DAY: M T W T F S Su
FOREMAN:	
CONTRACT No.:	

WEATHER CONDITIONS:		VISITORS:
AM	**PM**	
HOURS LOST DUE TO BAD WEATHER:		

SCHEDULE:		PROBLEMS/DELAYS:
Completion Date		
Days Ahead of Schedule		
Days Behind Schedule		

SAFETY ISSUES:	ACCIDENTS/INCIDENTS:

SUMMARY OF WORK PERFORMED TODAY

SIGNATURE	NAME

EMPLOYEE/ CONTRACTOR	TRADE	CONTRACTED HOURS	OVERTIME

EQUIPMENT ON SITE	NO. UNITS	Working	
		Yes	No

MATERIALS DELIVERED	NO. UNITS	EQUIPMENT RENTED	FROM & RATE

NOTES

_____

_____

_____

_____

DATE: / /	DAY: M T W T F S Su
FOREMAN:	
CONTRACT No.:	

WEATHER CONDITIONS:		VISITORS:
AM	**PM**	
HOURS LOST DUE TO BAD WEATHER:		

SCHEDULE:		PROBLEMS/DELAYS:
Completion Date		
Days Ahead of Schedule		
Days Behind Schedule		

SAFETY ISSUES:	ACCIDENTS/INCIDENTS:

SUMMARY OF WORK PERFORMED TODAY

SIGNATURE	NAME

EMPLOYEE/ CONTRACTOR	TRADE	CONTRACTED HOURS	OVERTIME

EQUIPMENT ON SITE	NO. UNITS	Working	
		Yes	No

MATERIALS DELIVERED	NO. UNITS	EQUIPMENT RENTED	FROM & RATE

NOTES

_____
_____
_____
_____

DATE: / /	DAY: M T W T F S Su
FOREMAN:	
CONTRACT No.:	

WEATHER CONDITIONS:		VISITORS:
AM	PM	
HOURS LOST DUE TO BAD WEATHER:		

SCHEDULE:		PROBLEMS/DELAYS:
Completion Date		
Days Ahead of Schedule		
Days Behind Schedule		

SAFETY ISSUES:	ACCIDENTS/INCIDENTS:

SUMMARY OF WORK PERFORMED TODAY

SIGNATURE	NAME

EMPLOYEE/ CONTRACTOR	TRADE	CONTRACTED HOURS	OVERTIME

EQUIPMENT ON SITE	NO. UNITS	Working	
		Yes	No

MATERIALS DELIVERED	NO. UNITS	EQUIPMENT RENTED	FROM & RATE

NOTES

_____

_____

_____

_____

DATE: / /	DAY: M T W T F S Su
FOREMAN:	
CONTRACT No.:	

WEATHER CONDITIONS:		VISITORS:
AM	**PM**	
HOURS LOST DUE TO BAD WEATHER:		

SCHEDULE:		PROBLEMS/DELAYS:
Completion Date		
Days Ahead of Schedule		
Days Behind Schedule		

SAFETY ISSUES:	ACCIDENTS/INCIDENTS:

SUMMARY OF WORK PERFORMED TODAY

SIGNATURE	NAME

EMPLOYEE/ CONTRACTOR	TRADE	CONTRACTED HOURS	OVERTIME

EQUIPMENT ON SITE	NO. UNITS	Working	
		Yes	No

MATERIALS DELIVERED	NO. UNITS	EQUIPMENT RENTED	FROM & RATE

NOTES

_____

_____

_____

_____

DATE: / /	DAY: M T W T F S Su

FOREMAN:	
CONTRACT No.:	

WEATHER CONDITIONS:		VISITORS:
AM	PM	
HOURS LOST DUE TO BAD WEATHER:		

SCHEDULE:		PROBLEMS/DELAYS:
Completion Date		
Days Ahead of Schedule		
Days Behind Schedule		

SAFETY ISSUES:	ACCIDENTS/INCIDENTS:

SUMMARY OF WORK PERFORMED TODAY

SIGNATURE	NAME

EMPLOYEE/ CONTRACTOR	TRADE	CONTRACTED HOURS	OVERTIME

EQUIPMENT ON SITE	NO. UNITS	Working	
		Yes	No

MATERIALS DELIVERED	NO. UNITS	EQUIPMENT RENTED	FROM & RATE

NOTES

_____
_____
_____
_____

DATE: / /	DAY: M T W T F S Su
FOREMAN:	
CONTRACT No.:	

WEATHER CONDITIONS:		VISITORS:
AM	**PM**	
HOURS LOST DUE TO BAD WEATHER:		

SCHEDULE:		PROBLEMS/DELAYS:
Completion Date		
Days Ahead of Schedule		
Days Behind Schedule		

SAFETY ISSUES:	ACCIDENTS/INCIDENTS:

SUMMARY OF WORK PERFORMED TODAY

SIGNATURE	NAME

EMPLOYEE/ CONTRACTOR	TRADE	CONTRACTED HOURS	OVERTIME

EQUIPMENT ON SITE	NO. UNITS	Working	
		Yes	No

MATERIALS DELIVERED	NO. UNITS	EQUIPMENT RENTED	FROM & RATE

NOTES

_____
_____
_____
_____

DATE: / /	DAY: M T W T F S Su
FOREMAN:	
CONTRACT No.:	

WEATHER CONDITIONS:		VISITORS:
AM	PM	
HOURS LOST DUE TO BAD WEATHER:		

SCHEDULE:		PROBLEMS/DELAYS:
Completion Date		
Days Ahead of Schedule		
Days Behind Schedule		

SAFETY ISSUES:	ACCIDENTS/INCIDENTS:

SUMMARY OF WORK PERFORMED TODAY

SIGNATURE	NAME

EMPLOYEE/ CONTRACTOR	TRADE	CONTRACTED HOURS	OVERTIME

EQUIPMENT ON SITE	NO. UNITS	Working	
		Yes	No

MATERIALS DELIVERED	NO. UNITS	EQUIPMENT RENTED	FROM & RATE

NOTES

DATE: / /	DAY: M T W T F S Su
FOREMAN:	
CONTRACT No.:	

WEATHER CONDITIONS:		VISITORS:
AM	**PM**	
HOURS LOST DUE TO BAD WEATHER:		

SCHEDULE:		PROBLEMS/DELAYS:
Completion Date		
Days Ahead of Schedule		
Days Behind Schedule		

SAFETY ISSUES:	ACCIDENTS/INCIDENTS:

SUMMARY OF WORK PERFORMED TODAY

SIGNATURE	NAME

EMPLOYEE/ CONTRACTOR	TRADE	CONTRACTED HOURS	OVERTIME

EQUIPMENT ON SITE	NO. UNITS	Working	
		Yes	No

MATERIALS DELIVERED	NO. UNITS	EQUIPMENT RENTED	FROM & RATE

NOTES

_____
_____
_____
_____

DATE: / /	DAY: M T W T F S Su
FOREMAN:	
CONTRACT No.:	

WEATHER CONDITIONS:		VISITORS:
AM	**PM**	
HOURS LOST DUE TO BAD WEATHER:		

SCHEDULE:		PROBLEMS/DELAYS:
Completion Date		
Days Ahead of Schedule		
Days Behind Schedule		

SAFETY ISSUES:	ACCIDENTS/INCIDENTS:

SUMMARY OF WORK PERFORMED TODAY

SIGNATURE	NAME

EMPLOYEE/ CONTRACTOR	TRADE	CONTRACTED HOURS	OVERTIME

EQUIPMENT ON SITE	NO. UNITS	Working	
		Yes	No

MATERIALS DELIVERED	NO. UNITS	EQUIPMENT RENTED	FROM & RATE

NOTES

_____
_____
_____
_____

DATE: / /	DAY: M T W T F S Su

FOREMAN:	
CONTRACT No.:	

WEATHER CONDITIONS:		VISITORS:
AM	PM	
HOURS LOST DUE TO BAD WEATHER:		

SCHEDULE:		PROBLEMS/DELAYS:
Completion Date		
Days Ahead of Schedule		
Days Behind Schedule		

SAFETY ISSUES:	ACCIDENTS/INCIDENTS:

SUMMARY OF WORK PERFORMED TODAY

SIGNATURE	NAME

EMPLOYEE/ CONTRACTOR	TRADE	CONTRACTED HOURS	OVERTIME

EQUIPMENT ON SITE	NO. UNITS	Working	
		Yes	No

MATERIALS DELIVERED	NO. UNITS	EQUIPMENT RENTED	FROM & RATE

NOTES

DATE: / /	DAY: M T W T F S Su
FOREMAN:	
CONTRACT No.:	

WEATHER CONDITIONS:		VISITORS:
AM	**PM**	
HOURS LOST DUE TO BAD WEATHER:		

SCHEDULE:		PROBLEMS/DELAYS:
Completion Date		
Days Ahead of Schedule		
Days Behind Schedule		

SAFETY ISSUES:	ACCIDENTS/INCIDENTS:

SUMMARY OF WORK PERFORMED TODAY

SIGNATURE	NAME

EMPLOYEE/ CONTRACTOR	TRADE	CONTRACTED HOURS	OVERTIME

EQUIPMENT ON SITE	NO. UNITS	Working	
		Yes	No

MATERIALS DELIVERED	NO. UNITS	EQUIPMENT RENTED	FROM & RATE

NOTES

DATE: / /	DAY: M T W T F S Su
FOREMAN:	
CONTRACT No.:	

WEATHER CONDITIONS:		VISITORS:
AM	PM	
HOURS LOST DUE TO BAD WEATHER:		

SCHEDULE:		PROBLEMS/DELAYS:
Completion Date		
Days Ahead of Schedule		
Days Behind Schedule		

SAFETY ISSUES:	ACCIDENTS/INCIDENTS:

SUMMARY OF WORK PERFORMED TODAY

SIGNATURE	NAME

EMPLOYEE/ CONTRACTOR	TRADE	CONTRACTED HOURS	OVERTIME

EQUIPMENT ON SITE	NO. UNITS	Working	
		Yes	No

MATERIALS DELIVERED	NO. UNITS	EQUIPMENT RENTED	FROM & RATE

NOTES

DATE: / /	DAY: M T W T F S Su

FOREMAN:	
CONTRACT No.:	

WEATHER CONDITIONS:		VISITORS:
AM	PM	
HOURS LOST DUE TO BAD WEATHER:		

SCHEDULE:		PROBLEMS/DELAYS:
Completion Date		
Days Ahead of Schedule		
Days Behind Schedule		

SAFETY ISSUES:	ACCIDENTS/INCIDENTS:

SUMMARY OF WORK PERFORMED TODAY

SIGNATURE	NAME

EMPLOYEE/ CONTRACTOR	TRADE	CONTRACTED HOURS	OVERTIME

EQUIPMENT ON SITE	NO. UNITS	Working	
		Yes	No

MATERIALS DELIVERED	NO. UNITS	EQUIPMENT RENTED	FROM & RATE

NOTES

_____

_____

_____

_____

DATE: / /	DAY: M T W T F S Su
FOREMAN:	
CONTRACT No.:	

WEATHER CONDITIONS:		VISITORS:
AM	**PM**	
HOURS LOST DUE TO BAD WEATHER:		

SCHEDULE:		PROBLEMS/DELAYS:
Completion Date		
Days Ahead of Schedule		
Days Behind Schedule		

SAFETY ISSUES:	ACCIDENTS/INCIDENTS:

SUMMARY OF WORK PERFORMED TODAY

SIGNATURE	NAME

EMPLOYEE/ CONTRACTOR	TRADE	CONTRACTED HOURS	OVERTIME

EQUIPMENT ON SITE	NO. UNITS	Working	
		Yes	No

MATERIALS DELIVERED	NO. UNITS	EQUIPMENT RENTED	FROM & RATE

NOTES

_____

_____

_____

_____

_____

DATE: / /	DAY: M T W T F S Su
FOREMAN:	
CONTRACT No.:	

WEATHER CONDITIONS:		VISITORS:
AM	PM	
HOURS LOST DUE TO BAD WEATHER:		

SCHEDULE:		PROBLEMS/DELAYS:
Completion Date		
Days Ahead of Schedule		
Days Behind Schedule		

SAFETY ISSUES:	ACCIDENTS/INCIDENTS:

SUMMARY OF WORK PERFORMED TODAY

SIGNATURE	NAME

EMPLOYEE/ CONTRACTOR	TRADE	CONTRACTED HOURS	OVERTIME

EQUIPMENT ON SITE	NO. UNITS	Working	
		Yes	No

MATERIALS DELIVERED	NO. UNITS	EQUIPMENT RENTED	FROM & RATE

NOTES

_____
_____
_____
_____

DATE: / /	DAY: M T W T F S Su
FOREMAN:	
CONTRACT No.:	

WEATHER CONDITIONS:		VISITORS:
AM	**PM**	
HOURS LOST DUE TO BAD WEATHER:		

SCHEDULE:		PROBLEMS/DELAYS:
Completion Date		
Days Ahead of Schedule		
Days Behind Schedule		

SAFETY ISSUES:	ACCIDENTS/INCIDENTS:

SUMMARY OF WORK PERFORMED TODAY

SIGNATURE	NAME

EMPLOYEE/ CONTRACTOR	TRADE	CONTRACTED HOURS	OVERTIME

EQUIPMENT ON SITE	NO. UNITS	Working	
		Yes	No

MATERIALS DELIVERED	NO. UNITS	EQUIPMENT RENTED	FROM & RATE

NOTES

DATE: / /	DAY: M T W T F S Su
FOREMAN:	
CONTRACT No.:	

WEATHER CONDITIONS:		VISITORS:
AM	PM	
HOURS LOST DUE TO BAD WEATHER:		

SCHEDULE:		PROBLEMS/DELAYS:
Completion Date		
Days Ahead of Schedule		
Days Behind Schedule		

SAFETY ISSUES:	ACCIDENTS/INCIDENTS:

SUMMARY OF WORK PERFORMED TODAY

SIGNATURE	NAME

EMPLOYEE/ CONTRACTOR	TRADE	CONTRACTED HOURS	OVERTIME

EQUIPMENT ON SITE	NO. UNITS	Working	
		Yes	No

MATERIALS DELIVERED	NO. UNITS	EQUIPMENT RENTED	FROM & RATE

NOTES

_____

_____

_____

_____

DATE: / /	DAY:	M	T	W	T	F	S	Su
FOREMAN:								
CONTRACT No.:								

WEATHER CONDITIONS:		VISITORS:
AM	**PM**	
HOURS LOST DUE TO BAD WEATHER:		

SCHEDULE:		PROBLEMS/DELAYS:
Completion Date		
Days Ahead of Schedule		
Days Behind Schedule		

SAFETY ISSUES:	ACCIDENTS/INCIDENTS:

SUMMARY OF WORK PERFORMED TODAY

SIGNATURE	NAME

EMPLOYEE/ CONTRACTOR	TRADE	CONTRACTED HOURS	OVERTIME

EQUIPMENT ON SITE	NO. UNITS	Working	
		Yes	No

MATERIALS DELIVERED	NO. UNITS	EQUIPMENT RENTED	FROM & RATE

NOTES

_____
_____
_____
_____

DATE: / /	DAY: M T W T F S Su
FOREMAN:	
CONTRACT No.:	

WEATHER CONDITIONS:		VISITORS:
AM	**PM**	
HOURS LOST DUE TO BAD WEATHER:		

SCHEDULE:		PROBLEMS/DELAYS:
Completion Date		
Days Ahead of Schedule		
Days Behind Schedule		

SAFETY ISSUES:	ACCIDENTS/INCIDENTS:

SUMMARY OF WORK PERFORMED TODAY

SIGNATURE	NAME

EMPLOYEE/ CONTRACTOR	TRADE	CONTRACTED HOURS	OVERTIME

EQUIPMENT ON SITE	NO. UNITS	Working	
		Yes	No

MATERIALS DELIVERED	NO. UNITS	EQUIPMENT RENTED	FROM & RATE

NOTES

DATE: / /	DAY: M T W T F S Su

FOREMAN:	

CONTRACT No.:	

WEATHER CONDITIONS:		VISITORS:
AM	PM	
HOURS LOST DUE TO BAD WEATHER:		

SCHEDULE:		PROBLEMS/DELAYS:
Completion Date		
Days Ahead of Schedule		
Days Behind Schedule		

SAFETY ISSUES:	ACCIDENTS/INCIDENTS:

SUMMARY OF WORK PERFORMED TODAY

SIGNATURE	NAME

EMPLOYEE/ CONTRACTOR	TRADE	CONTRACTED HOURS	OVERTIME

EQUIPMENT ON SITE	NO. UNITS	Working	
		Yes	No

MATERIALS DELIVERED	NO. UNITS	EQUIPMENT RENTED	FROM & RATE

NOTES

DATE: / /	DAY: M T W T F S Su
FOREMAN:	
CONTRACT No.:	

WEATHER CONDITIONS:		VISITORS:
AM	**PM**	
HOURS LOST DUE TO BAD WEATHER:		

SCHEDULE:		PROBLEMS/DELAYS:
Completion Date		
Days Ahead of Schedule		
Days Behind Schedule		

SAFETY ISSUES:	ACCIDENTS/INCIDENTS:

SUMMARY OF WORK PERFORMED TODAY

SIGNATURE	NAME

EMPLOYEE/ CONTRACTOR	TRADE	CONTRACTED HOURS	OVERTIME

EQUIPMENT ON SITE	NO. UNITS	Working	
		Yes	No

MATERIALS DELIVERED	NO. UNITS	EQUIPMENT RENTED	FROM & RATE

NOTES

_____

_____

_____

_____

DATE:	/	/	DAY:	M	T	W	T	F	S	Su
FOREMAN:										
CONTRACT No.:										

WEATHER CONDITIONS:		VISITORS:
AM	**PM**	
HOURS LOST DUE TO BAD WEATHER:		

SCHEDULE:		PROBLEMS/DELAYS:
Completion Date		
Days Ahead of Schedule		
Days Behind Schedule		

SAFETY ISSUES:	ACCIDENTS/INCIDENTS:

SUMMARY OF WORK PERFORMED TODAY

SIGNATURE	NAME

EMPLOYEE/ CONTRACTOR	TRADE	CONTRACTED HOURS	OVERTIME

EQUIPMENT ON SITE	NO. UNITS	Working	
		Yes	No

MATERIALS DELIVERED	NO. UNITS	EQUIPMENT RENTED	FROM & RATE

NOTES

_____

_____

_____

_____

DATE:	/	/	DAY:	M	T	W	T	F	S	Su
FOREMAN:										
CONTRACT No.:										

WEATHER CONDITIONS:		VISITORS:
AM	**PM**	
HOURS LOST DUE TO BAD WEATHER:		

SCHEDULE:		PROBLEMS/DELAYS:
Completion Date		
Days Ahead of Schedule		
Days Behind Schedule		

SAFETY ISSUES:	ACCIDENTS/INCIDENTS:

SUMMARY OF WORK PERFORMED TODAY

SIGNATURE	NAME

EMPLOYEE/ CONTRACTOR	TRADE	CONTRACTED HOURS	OVERTIME

EQUIPMENT ON SITE	NO. UNITS	Working	
		Yes	No

MATERIALS DELIVERED	NO. UNITS	EQUIPMENT RENTED	FROM & RATE

NOTES

_____
_____
_____
_____

DATE: / /	DAY: M T W T F S Su
FOREMAN:	
CONTRACT No.:	

WEATHER CONDITIONS:		VISITORS:
AM	**PM**	
HOURS LOST DUE TO BAD WEATHER:		

SCHEDULE:		PROBLEMS/DELAYS:
Completion Date		
Days Ahead of Schedule		
Days Behind Schedule		

SAFETY ISSUES:	ACCIDENTS/INCIDENTS:

SUMMARY OF WORK PERFORMED TODAY

SIGNATURE	**NAME**

EMPLOYEE/ CONTRACTOR	TRADE	CONTRACTED HOURS	OVERTIME

EQUIPMENT ON SITE	NO. UNITS	Working	
		Yes	No

MATERIALS DELIVERED	NO. UNITS	EQUIPMENT RENTED	FROM & RATE

NOTES

DATE: / /	DAY: M T W T F S Su
FOREMAN:	
CONTRACT No.:	

WEATHER CONDITIONS:		VISITORS:
AM	PM	
HOURS LOST DUE TO BAD WEATHER:		

SCHEDULE:		PROBLEMS/DELAYS:
Completion Date		
Days Ahead of Schedule		
Days Behind Schedule		

SAFETY ISSUES:	ACCIDENTS/INCIDENTS:

SUMMARY OF WORK PERFORMED TODAY

SIGNATURE	NAME

EMPLOYEE/ CONTRACTOR	TRADE	CONTRACTED HOURS	OVERTIME

EQUIPMENT ON SITE	NO. UNITS	Working	
		Yes	No

MATERIALS DELIVERED	NO. UNITS	EQUIPMENT RENTED	FROM & RATE

NOTES

_____

_____

_____

_____

| DATE: / / | DAY: M T W T F S Su |

| FOREMAN: | |

| CONTRACT No.: | |

WEATHER CONDITIONS:	VISITORS:
AM	PM	

HOURS LOST DUE TO BAD WEATHER:

SCHEDULE:	PROBLEMS/DELAYS:
Completion Date	
Days Ahead of Schedule	
Days Behind Schedule	

SAFETY ISSUES:	ACCIDENTS/INCIDENTS:

SUMMARY OF WORK PERFORMED TODAY

SIGNATURE	NAME

EMPLOYEE/ CONTRACTOR	TRADE	CONTRACTED HOURS	OVERTIME

EQUIPMENT ON SITE	NO. UNITS	Working	
		Yes	No

MATERIALS DELIVERED	NO. UNITS	EQUIPMENT RENTED	FROM & RATE

NOTES

_____

_____

_____

_____

DATE:	/	/	DAY:	M	T	W	T	F	S	Su
FOREMAN:										
CONTRACT No.:										

WEATHER CONDITIONS:		VISITORS:
AM	**PM**	
HOURS LOST DUE TO BAD WEATHER:		

SCHEDULE:		PROBLEMS/DELAYS:
Completion Date		
Days Ahead of Schedule		
Days Behind Schedule		

SAFETY ISSUES:	ACCIDENTS/INCIDENTS:

SUMMARY OF WORK PERFORMED TODAY

SIGNATURE	NAME

EMPLOYEE/ CONTRACTOR	TRADE	CONTRACTED HOURS	OVERTIME

EQUIPMENT ON SITE	NO. UNITS	Working	
		Yes	No

MATERIALS DELIVERED	NO. UNITS	EQUIPMENT RENTED	FROM & RATE

NOTES

DATE:	/	/	DAY:	M	T	W	T	F	S	Su
FOREMAN:										
CONTRACT No.:										

WEATHER CONDITIONS:		VISITORS:
AM	**PM**	
HOURS LOST DUE TO BAD WEATHER:		

SCHEDULE:		PROBLEMS/DELAYS:
Completion Date		
Days Ahead of Schedule		
Days Behind Schedule		

SAFETY ISSUES:	ACCIDENTS/INCIDENTS:

SUMMARY OF WORK PERFORMED TODAY

SIGNATURE	NAME

EMPLOYEE/ CONTRACTOR	TRADE	CONTRACTED HOURS	OVERTIME

EQUIPMENT ON SITE	NO. UNITS	Working	
		Yes	No

MATERIALS DELIVERED	NO. UNITS	EQUIPMENT RENTED	FROM & RATE

NOTES

DATE: / /	DAY: M T W T F S Su
FOREMAN:	
CONTRACT No.:	

WEATHER CONDITIONS:		VISITORS:
AM	**PM**	
HOURS LOST DUE TO BAD WEATHER:		

SCHEDULE:		PROBLEMS/DELAYS:
Completion Date		
Days Ahead of Schedule		
Days Behind Schedule		

SAFETY ISSUES:	ACCIDENTS/INCIDENTS:

SUMMARY OF WORK PERFORMED TODAY

SIGNATURE	NAME

EMPLOYEE/ CONTRACTOR	TRADE	CONTRACTED HOURS	OVERTIME

EQUIPMENT ON SITE	NO. UNITS	Working	
		Yes	No

MATERIALS DELIVERED	NO. UNITS	EQUIPMENT RENTED	FROM & RATE

NOTES

_____

_____

_____

_____

DATE: / /	DAY: M T W T F S Su
FOREMAN:	
CONTRACT No.:	

WEATHER CONDITIONS:		VISITORS:
AM	**PM**	
HOURS LOST DUE TO BAD WEATHER:		

SCHEDULE:		PROBLEMS/DELAYS:
Completion Date		
Days Ahead of Schedule		
Days Behind Schedule		

SAFETY ISSUES:	ACCIDENTS/INCIDENTS:

SUMMARY OF WORK PERFORMED TODAY

SIGNATURE	NAME

EMPLOYEE/ CONTRACTOR	TRADE	CONTRACTED HOURS	OVERTIME

EQUIPMENT ON SITE	NO. UNITS	Working	
		Yes	No

MATERIALS DELIVERED	NO. UNITS	EQUIPMENT RENTED	FROM & RATE

NOTES

_____

_____

_____

_____

DATE: / /	DAY: M T W T F S Su
FOREMAN:	
CONTRACT No.:	

WEATHER CONDITIONS:		VISITORS:
AM	**PM**	
HOURS LOST DUE TO BAD WEATHER:		

SCHEDULE:		PROBLEMS/DELAYS:
Completion Date		
Days Ahead of Schedule		
Days Behind Schedule		

SAFETY ISSUES:	ACCIDENTS/INCIDENTS:

SUMMARY OF WORK PERFORMED TODAY

SIGNATURE	NAME

EMPLOYEE/ CONTRACTOR	TRADE	CONTRACTED HOURS	OVERTIME

EQUIPMENT ON SITE	NO. UNITS	Working	
		Yes	No

MATERIALS DELIVERED	NO. UNITS	EQUIPMENT RENTED	FROM & RATE

NOTES

Made in the USA
Middletown, DE
04 October 2023

40237144R00068